I0606269

STEM CAREER CHOICES

How to Choose Your Perfect Math Career

Cathleen Small

CHERITON
CHILDREN'S BOOKS

Published in 2023 by **Cheriton Children's Books**
PO Box 7258, Bridgnorth, Shropshire, WV16 9ET, UK

First Edition

Author: Cathleen Small
Designer: Paul Myerscough
Editors: Sarah Eason and Jennifer Sanderson
Proofreader: Ella Short

Picture credits: Cover: Shutterstock/Stockfour. Inside: pp. 1, 13, 19: Shutterstock/Burben; pp. 1, 22: Shutterstock/Gorodenkoff; p. 4: Shutterstock/BigBlues; p. 5: Shutterstock/NicoElNino; pp. 6, 12, 19: Shutterstock/Drazen Zigic; pp. 6, 21, 28: Shutterstock/Chad McDermott; pp. 6, 35: Shutterstock/ESB Professional; pp. 7, 42: Shutterstock/Portrait Image Asia; pp. 7, 54, 59: Shutterstock/Elnur; p. 8: Shutterstock/BalanceFormCreative; p. 9: Shutterstock/Rido; pp. 10, 18: Shutterstock/Goodluz; pp. 11, 18: Shutterstock/Sirichai Saengcharnchai; p. 14: Shutterstock/Bernhard Staehli; p. 15: Shutterstock/Everett Collection; p. 16: Shutterstock/Michaeljung; p. 17: Shutterstock/Michaeljung; pp. 20, 28: Shutterstock/Indypendenz; pp. 23, 29: Shutterstock/Cristovao; pp. 24, 29: Shutterstock/Alexander Gatsenko; p. 25: Shutterstock/Ground Picture; p. 26: Shutterstock/Rido; p. 27: Shutterstock/Rido; p. 30: Shutterstock/Frame Stock Footage; pp. 31, 38: Shutterstock/Fizkes; p. 32: Shutterstock/Fizkes; pp. 33, 38: Shutterstock/GaudiLab; pp. 34, 39: Shutterstock/Francesco Scatena; p. 36: Shutterstock/Frame Stock Footage; p. 37: Shutterstock/Frame Stock Footage; p. 39: Shutterstock/Yakobchuk Viacheslav; pp. 40, 48: Shutterstock/Roman Samborskyi; pp. 41, 48: Shutterstock/Rawpixel.com; p. 43: Shutterstock/Sasirin Pamai; pp. 44, 49: Shutterstock/Standret; p. 45: Shutterstock/Monkey Business Images; pp. 46, 49: Shutterstock/Andrey Popov; p. 47: Shutterstock/Andrey Popov; pp. 50, 58: Shutterstock/Baranq; pp. 51, 59: Shutterstock/Adam Gregor; p. 52: Shutterstock/Fizkes; p. 53: Shutterstock/Massimo Parisi; p. 55: Shutterstock/Kathy Hutchins; pp. 56, 58: Shutterstock/Roman Samborskyi; p. 57: Shutterstock/Roman Samborskyi; p. 60: Shutterstock/Fizkes; p. 61: Shutterstock/Asier Romero.

Printed in China

CONTENTS

CHAPTER ONE
WHO DO YOU THINK YOU ARE? 4

CHAPTER TWO
HELPER ROLES IN MATH 10

CHAPTER THREE
BUILDER ROLES IN MATH 20

CHAPTER FOUR
CREATOR ROLES IN MATH 30

CHAPTER FIVE
ORGANIZER ROLES IN MATH 40

CHAPTER SIX
THINKER ROLES IN MATH 50

WHAT NEXT?–YOUR CAREER CHECKLIST 60

GLOSSARY 62
FIND OUT MORE 63
INDEX AND ABOUT THE AUTHOR 64

CHAPTER ONE

WHO DO YOU THINK YOU ARE?

If you enjoy math and you're good at it, too, you may be thinking about a career in math. This is great news because in this vibrant career area, there are many options to explore.

Finding Your Way

Math is a really big field, with many careers, from biostatistician or air traffic controller to architect or urban planner. There are many, many options to choose from.

You might assume that you need a college degree to have a career in math. There are definitely careers where you will need a degree in math to find work but college isn't for everyone, and there are math careers you can get into that don't necessarily require a four-year degree. Whatever your reasons for not attending college, you don't have to completely give up on the idea of having a career in math. You have options to explore.

Reading a book about math careers to try to narrow down what you think you might like to do is a good place to start. This book will help you determine your personality type and then explore careers in math that usually appeal to people with personalities like yours. So let's get started!

If you are struggling to find a career that you think will suit you, personality testing can help kick-start that thought process.

Finding a career that makes the most of your talents and in which you feel happy and satisfied is a truly great outcome.

Follow the Flowchart to Fast-Track Your Future!

To find your personality type, you'll need to work through the flowchart on the following pages. The flowchart asks you simple questions to help you determine your personality type. This is based on five types of personality and the work situations that typically appeal to people with those types of personality. The five personality types in the flowchart are: social; practical; creative; organized; and analytical.

These personality types correspond to five different work environments. Social people, for example, often enjoy helper types of careers. Practical people often thrive in builder types of jobs. Creative people flourish in creator roles. Organized people typically enjoy organizational roles. And analytical people usually gravitate to thinking jobs. Although flowcharts are not foolproof, the one in this book should help steer you in the right direction.

Exploring your personality type and career options is important because if you work in an area that is a good fit for you, you will be more likely to be satisfied and successful. And that, in turn, will lead to greater overall happiness with your life. So take a look at the flowchart on the next pages and work through it to determine your career personality type. Then, follow the next steps to find your perfect career in math.

This flowchart asks you questions about your preferences to help you figure out which of the five personality types best describes you. It helps you think about what you like and don't like, and what kind of work might be best for you, so you can make sound career choices.

Once you have figured out what your career type is, take a look at the career choices in this book. Each chapter features some interesting careers linked to the personality types shown in the flowchart on these pages. A variety of jobs are explored in each chapter, along with a day in the life of one of the roles featured. Each chapter concludes with a checklist that helps you work through how you feel about the featured jobs and if they may be right for you.

Are you interested in helping people and the environment?

No, this is not an area I want to work in

Yes, I want to make a positive contribution to people's lives and the environment

Helper

Do you like working in a practical, hands-on way?

No, I hate having to be practical

Yup, I am super practical

Builder

WHAT'S YOUR CAREER TYPE?

Creator

Organizer

Are you artistic or creative?

That's me! → Creator

No, art's not my thing! → Do you love having things in order?

Do you love having things in order?

Yes! Check that box! → Organizer

No, I don't care about that → Do you enjoy studying and thinking through complex ideas?

Do you enjoy studying and thinking through complex ideas?

Yes, I love theories and thinking through ideas → Thinker

No, I don't enjoy complicated study → If you have made it all the way to this box, try taking the test again–you can work through it a few times before you finalize your answer.

Thinker

What Did You Learn?

Did you discover anything new about yourself from the flowchart? Did you find a career personality that fits you like a glove or did you find it hard to answer the questions on the chart and narrow it down to just one personality type?

If you found it difficult to narrow down your work personality type to one, that's actually a good thing: It means you have a wider range of job areas to look at. Most people find they fit more than one type of personality: Our personalities are made up of many different parts, so you'll often find that more than one personality type fits you.

Finding a good career fit is a little like solving a math problem—you need to work through each stage, check your answers, and come to a conclusion that best solves the problem.

Weed Out the Nos

If you had a hard time narrowing down the personality type that best fits you, one way to move forward is by weeding out the personality types that definitely do not fit you. For example, if you hate getting your hands dirty, a builder-type job probably isn't the best place to start your career search. So weed out the nos first, and then review the remaining personality types on the flowchart and make those your starting point. And after that?...

What to Do with the Yesses

Start with all the yesses. Take your time to work through the book and learn more about the possible math careers that are best suited to your personality type or types. Each chapter of this book focuses on specific math careers that often appeal to people who fit into the given personality type.

Math has so many different career options that if you start off in one area, such as architecture, and discover that you don't like it, you can retrain and still use your math skills in another area, such as teaching schoolchildren.

Although you should start with the chapters that cover the career personality types that best seem to fit you, don't miss the other chapters you passed on earlier. You never know when a particular career might resonate with you, even if it doesn't seem a good match at first glance.

Explore Options

This book covers many options in the field of math but not all of them. You can read the book to get started and then explore further. The Review and Check In sections at the end of each chapter feature further career options you could research. And once you reach the end of the book, the What Next? checklist will guide you through taking the next steps to kick-start your career.

Career Insight: Nothing Is Ever Set in Stone

Always remember that nothing is ever permanent. If you enter a career and find you do not enjoy it, it is not the end of the world. You can certainly switch to another career in the field or even switch to an entirely different field. The important thing is to keep checking in with yourself and make sure you are on a path that feels right for you.

CHAPTER TWO

HELPER ROLES IN MATH

If you're a helper, your goals probably involve helping people and making a contribution to the greater good. Although math may seem like a solitary activity, there are actually a lot of ways you can use math to help people and the environment. So, how can you put your skills to good use?

Biostatistician

Statisticians prepare and analyze numerical data to draw conclusions about what that data represents. Biostatisticians work specifically on data about human health, plants, and animals. It's a career that marries statistics and biology.

Biostatisticians often work in healthcare and pharmaceutical fields or for government agencies. The data they work with often relates to disease and genetics. When looking at data, it's important to know if something is statistically significant. For example, suppose you have been tasked to look at the genetic makeup of 1,000 people who are very talented at music. If you found a specific genetic code in 2 of these people, that would not be statistically significant because 2 people out of 1,000 is a very, very small percentage. However, what if that specific genetic code was found in 750 of the 1,000 people who are very talented at music? Now that might be statistically significant—maybe you found the genetic key to what makes people talented musicians. These are the kinds of patterns biostatisticians look at. When studying biology, genetics, diseases, or conditions, they look at data and try to find where the patterns are. If there is a statistically significant pattern, they may have helped unlock the mystery of what they're studying and may be able to help millions of people in the process.

If you love analyzing numbers and other data, becoming a biostatistician could be a great career path to follow.

Epidemiologists are at the cutting edge of work to prevent disease transmission. They carefully study viruses and pathogens that cause diseases, and use that information to advise people how best to protect themselves.

To become a biostatistician, you will need a degree in mathematics, statistics, or a related field. Some employers will hire those with just a bachelor's degree but many employers prefer to hire biostatisticians with a master's degree or PhD in the field.

Epidemiologist

Another career in the health field that requires a strong background in math is that of an epidemiologist. Epidemiologists study what causes diseases and how they spread, and then develop plans for how to prevent further spread of the diseases.

During the Covid-19 pandemic, epidemiologists were on the front line of information. Throughout the pandemic, epidemiologists' analysis of data on infections and deaths, and projections from studies that model the virus's spread, drove government policy decisions all over the world. Many of these, such as locking down countries, imposing quarantines, and enforcing social distancing and mask-wearing, became commonplace, thanks to the data worked on by epidemiologists.

To be an epidemiologist, you must have at least a master's degree in public health (MPH) or a related field. Many epidemiologists also have a doctoral degree in either epidemiology or medicine.

Math Teacher

One great way to help others is by teaching the next generation of mathematicians. If you enjoy math and you like working with young people, then a teaching career might be for you. If you want to teach math specifically (rather than teaching all subjects), you'll need to teach at middle school or high school.

Middle school and high school teachers must have a bachelor's degree. Some schools prefer to hire teachers with a master's degree but in many cases, a bachelor's degree is enough. If you are teaching in a public school, you'll also need a state-issued certification or licence to teach.

Air Traffic Controller

It might surprise you to learn that one way you can use math to help people is by being an air traffic controller. Air traffic controllers track dozens of planes in the sky at once, all on a small screen. They must be able to quickly do math in their mind as they compute the distances between planes and determine what altitude each pilot should be flying at.

There are a number of ways to become an air traffic controller. The best way is to earn a degree through the Air Traffic-Collegiate Training Initiative (AT-CTI). There are two-year and four-year degrees available but if you choose the

Math is a very important subject for schoolchildren of all ages today. Maybe you could help teach them.

Air traffic controllers must use their skills in computing mathematical information to make quick decisions that keep the skies safe.

shorter degree, you must make up for the extra years by having relevant work experience in the field.

Alternatively, you can earn a bachelor's degree in a field such as math or engineering instead of going through AT-CTI but you'll have to take a five-week Air Traffic Basics course at the Federal Aviation Administration (FAA) Academy. That five-week course is in addition to the rest of the FAA Academy training that air traffic controllers must go through.

Career Insight:
Forensic Science Technician

Another math-related career you might not have thought of is a forensic science technician. They analyze data at crime scenes to help solve crimes and determine exactly what happened at a particular scene. Part of their data analysis involves recording the position and location of found evidence and drawing conclusions about it. For example, a forensic science technician might need to determine what sort of tool was used by a burglar to smash a window and break into a house. They would measure how far broken glass from the window had spread into the house. A heavier tool would bring more force, which would spread bits of glass much farther inside the house than if the burglar had used their fist or a small tool. If that kind of detective work appeals to you, you might explore a job as a forensic science technician.

Climate change is one of the greatest challenges of modern times. If you choose to use your math skills as a climatologist you could contribute to finding out more about how climate change is likely to affect us in the future.

Climatologist

If you're a helper but your interest in helping is more geared toward environmental work rather than helping people directly, you might be interested in a career in climatology. Climatology is the study of climate and how it changes over time. Climatology is not to be confused with meteorology, which looks at short-term weather patterns and events.

Climatologists collect data to make predictions about future climate patterns. They then communicate their research findings to the international scientific community and the public. Often, their findings are used to guide people to more sustainable living.

Some of a climatologist's duties may include analyzing and interpreting data that is obtained from meteorological stations, radar and satellite imagery, and computer models. They may also examine historical data and then use the information together with recent data to predict future climate trends. Climatologists also conduct research on atmospheric events, such as thunderstorms, hurricanes, and cyclones, and use this information to predict future events and their severity. They also look at how climate affects biomes, such as wetlands, and human health. During a time when climate change is a concern to many, climatology is an exciting field to work in.

Many climatologists work for the government as part of the National Weather Service (NWS), providing forecasts, warnings of hazardous weather, and other weather-related events. Others work for private scientific and technical firms. Entry level climatology positions require a bachelor's degree in climatology or a related field, such as meteorology or atmospheric science. For research and academic positions, a master's or PhD will be required.

Career Insight:
The Nurse Mathematician

The name Florence Nightingale may be familiar to many as the founder of modern nursing. However, Nightingale is less known as a brilliant mathematician. Nightingale used her passion for math and helping people, and put them together in a career that ended up improving public health for all patients. While working as a nurse in the Crimean War, Nightingale was horrified by hospital conditions. She began gathering data about how unsanitary hospital conditions were leading to patient deaths from contagious diseases such as typhoid and cholera. She wrote a report that included statistical graphs and charts in an effort to prove her point. One type of graph Nightingale used is what we now call the pie chart. Pie charts are used often in reports from almost every career field. Nightingale's data was presented so convincingly to the government that it began to improve healthcare conditions. Her information led to better patient care and fewer deaths.

Florence Nightingale was a pioneer in more ways than one: as a nurse on the battlefield and as a female mathematician.

A DAY IN THE LIFE OF A High School Math Teacher

Teaching can be one of the most rewarding jobs but it requires a lot of effort, patience, and great commitment. Take a look at this day in the life of a high school math teacher to discover more.

6:00 a.m. School starts early, so your workday begins early too. You have to be at school before the day starts so that you can catch up on administration work. There are also sometimes staff meetings before school starts.

7:30 a.m. You arrive on campus and go straight to your classroom. You answer emails from your department head and there's also a message from a parent whose child is not doing especially well in your class. You let the parent know that you're aware of the problem and will think of a way forward to help the student.

8:00 a.m. Your first period is an algebra class. Your students have a test today, so you get to start your day slowly and quietly, while the students work hard on the test.

9:00 a.m. Your second-period geometry class arrives. You're teaching a lesson on finding the volume of three-dimensional (3-D) cylinders and spheres today. Working with curved objects is always a challenge for some students, but by the end of the lesson, most of them seem confident in their work.

Your classes are made up of mixed abilities, so one of your main tasks is to ensure everyone follows the information you have written on the board.

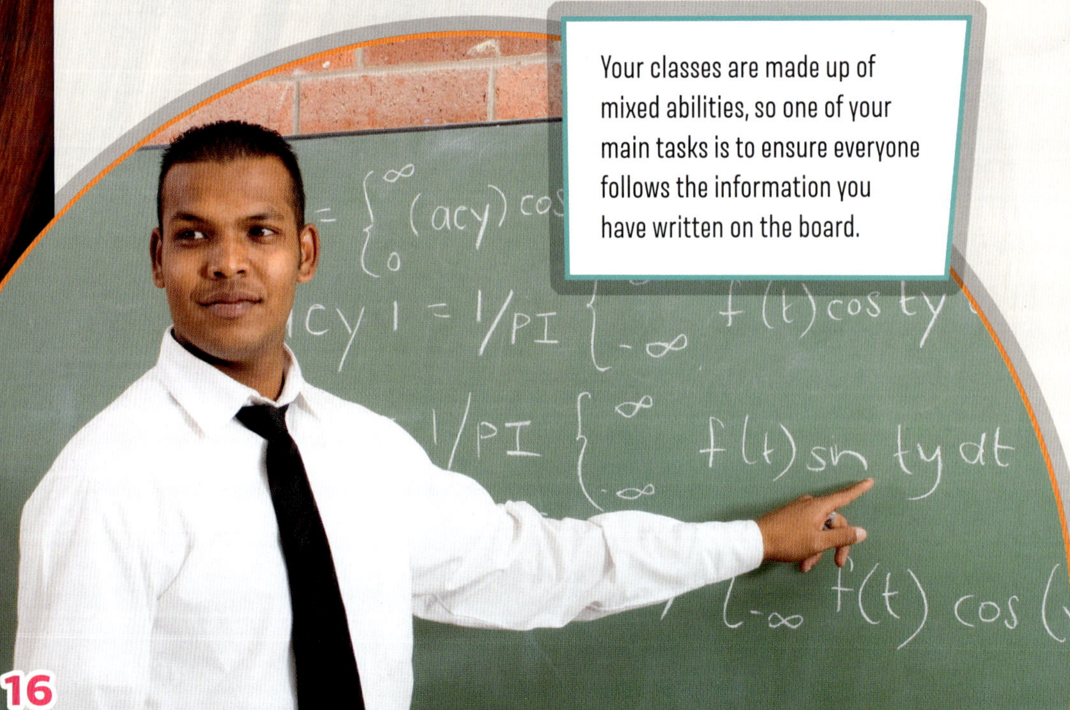

10:00 a.m. Your third-period class is also taking an algebra test. While your students work, you start grading the first period's tests.

11:00 a.m. Your fourth period is an AP calculus class. This is your favorite class to teach because it's the highest-level class offered at the school, and the students are really motivated. These are the students who really enjoy math, and they're a lot of fun to teach as a result.

It is always satisfying to help students figure out math problems–getting your class to engage and enjoy the subject feels great.

12:00 p.m. Being a teacher means that your lunch is part of the timetable, but sometimes there are meetings over lunch or you have to help a student with their work.

1:00 p.m. Your first class after lunch is a trigonometry class. Your students are working on sine and cosine, and several need extra help. You have a teaching assistant for that class, so you let half the class work independently while you lead a small-group review for the students who need some extra help.

2:00 p.m. In the sixth period, you're teaching the same lesson on 3-D cylinders and spheres that the second-period class did. Although you've taught this lesson before there are always different questions and problems, so no two lessons are ever the same.

3:00 p.m. Your seventh period pre-algebra class comes in. This is a remedial class for students who didn't pass the test to start high school in algebra. It's a smaller class because the students need extra attention, so you do a lot of small-group work. Thankfully, you have a teaching assistant for this period too, so you're able to give the students individual attention.

3:55 p.m. School is dismissed but your day isn't over yet. You always stay for at least an hour after school in case students have questions. That also lets you get some grading done, which reduces the work you have to take home.

Helper

Review and Check In

Now that you've learned about some of the helper roles in math, explore more about how you feel about each job. Use the questions below to help you assess how strongly you feel each role might suit you.

Biostatistician

- Are you happy and motivated to pursue the further education required to be a biostatistician?
- Is there anything in particular about the job that fascinates you?
- Is there anything that you do not like about the role? What is that? Do the benefits of the job outweigh the aspects you may not like?
- If you could choose to work in healthcare, pharmaceutics, or for the government, which one would you choose, and why?
- What skills do you have that make you well suited to do this job?

Epidemiologist

- What personality traits and skills do you have that make you well suited to do this job?
- You are likely to work as part of a team, collaborating with other scientists. Is this something you would enjoy? Why do you say so?
- You will need good communication skills to present and discuss your research findings. Is this something you would be comfortable with? If not, how could you improve on your communication skills?
- How might working on the front line of a crisis, such as a pandemic, affect you?

"Always keep in mind what you want from your career."

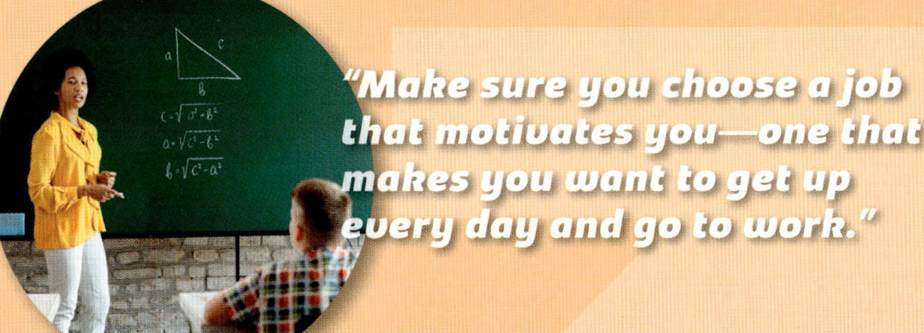

"Make sure you choose a job that motivates you—one that makes you want to get up every day and go to work."

Climatologist

- Why would you want to do this job?
- This job may require you to go into the field to collect data. Would you enjoy doing this? Why or why not?
- What skills do you have that make you well suited to this job?

Math Teacher

- Teaching is a very rewarding job. What else do you find attractive about being a teacher?
- You'll need a lot of patience to do this job, what other personality traits would help make you a good teacher?

Air Traffic Controller

- You would be responsible for hundreds of lives. What skills do you have that would help you cope with that pressure?
- Would doing shift work be a challenge for you? How do you think you could overcome this challenge?

"Always think, 'What am I good at?' then find a career that suits your skills."

Research More

If you like the idea of working as a helper in math, but are not sure the jobs covered in this chapter suit you, here are some more helper roles you could explore.

Finance Systems Support Analyst
Mathematical Healthcare Modeler
Anesthetist

CHAPTER THREE

BUILDER ROLES IN MATH

If you like to work with your hands and contribute to finishing tangible products, there are a number of math careers that typically appeal to builder-type personalities. Math is a skill that is required in many practical roles in which precise measurements and calculations are used.

Architect

Architects design buildings, houses, and other structures. They aren't the people pounding the nails in at the building site but they're the ones who design the blueprints that act as a guide for the construction team. Their ideas are put into practice.

Architects use math in many ways. They have to use geometry when designing buildings, and they have to do countless mathematical computations to make sure everything fits together in the structure as planned. Safety is a huge issue when designing structures, so architects need to be able to perform data calculations to make sure that everything is built safely. For example, if a beam will only hold a certain amount of weight, the architect must perform many calculations to ensure that whatever is being put on top of the beam will not be too heavy. Architects are often in charge of giving estimates on construction time and costs, which requires calculations. They also have to manage construction contracts, which also involve calculating costs.

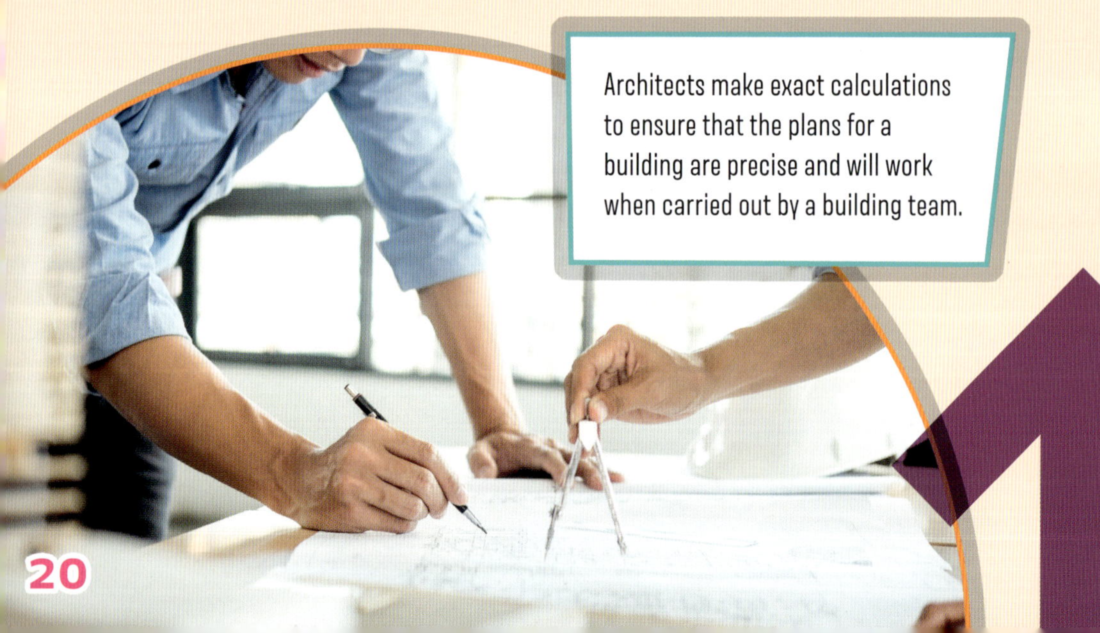

Architects make exact calculations to ensure that the plans for a building are precise and will work when carried out by a building team.

A lot of an architect's work can be done on computers but that doesn't mean an architect does not have to have mathematical knowledge. A strong knowledge of math is a must for this job. To become an architect, you must earn at least a bachelor's degree in architecture, work in the field at a paid internship, and then also pass the Architect Registration Examination. Many architectural firms prefer candidates who have a master's in architecture or are interested in earning a master's in the future.

Urban Planner

If building intrigues you but you'd like to do it on a larger scale than an architect does, what about a career in urban planning? Urban planners develop land-use plans for urban, suburban, and rural communities. This is important when communities are growing or being redeveloped, because as more people arrive in an area additional roads, schools, and other structures are needed. Urban planners help plan for that growth.

Just as architects need a solid math foundation, especially in geometry, to be able to design homes and other buildings, so too do urban planners. Public planning is a lot like planning a house: You have a certain amount of space and a certain number of things that need to fit into that space, and math will help you figure out how to best fit it all.

Urban planners often make to-scale models of the areas that they are planning in a town or city to help them see their plans in a 3-D form.

The Town's Needs

Urban planners also try to predict a community's future needs by analyzing data through mathematical models. For example, what does doubling the city's population mean for the number of grocery stores needed per square mile? How many schools will be needed in the future? Is there space to build a new school and where would be the best place for it to be built? Mathematical models can help determine answers to questions such as these.

Usually, urban planners must have a master's degree in urban planning to work in the field. There are a number of good bachelor's degree areas for those just entering college and beginning their work toward a career in urban planning. Economics, geography, political science, and environmental design are all recommended bachelor's programs for students interested in eventually working in urban planning. Most states do not require urban planners to be certified to practice.

Mechanical Engineer

If you're good at math and you enjoy hands-on, practical types of jobs, then mechanical engineering is a field that may interest you. Mechanical engineers design, build, and test mechanical devices, such as engines and machines. To design and build devices, mechanical engineers need to have a strong understanding of mathematical relations. Computers may perform the actual calculations but computer programs are only usable if you understand what they are doing and what numbers they need to be able to calculate.

To become a mechanical engineer, you'll need at least a bachelor's degree in mechanical engineering or a related field. Some employers prefer their employees to have a master's degree in the field.

For an entry-level engineering position, you don't need to be licensed. However, if you want to advance to higher positions, you'll need to train further and earn your Professional Engineering (PE) license. To earn the license, you'll need four or more years of experience in your field and pass scores on the Fundamentals of Engineering (FE) and Professional Engineering exam.

Today, many more women are entering the world of engineering.

Engineering has long been a field dominated by men. Female engineers exist, but they are few and far between. According to a recent study by the Bureau of Labor Statistics (BLS), only 14 percent of engineers in the workforce are women. However, there are some female engineers who are doing really amazing things. For example, Meredith Westafer, an industrial engineering manager for Tesla is managing the design and layout of Tesla's massive new factory in Nevada. Westafer used mathematical models to determine how much space was needed for the factory and how the workflow will exist within that space. She even designed a system in which robots will deliver materials to the production line.

Surveyor or Mapping Technician

You might have noticed that, so far, all of the building careers in math we have discussed require a college degree. However, there are a few related jobs that do not require advanced education. For example, surveying technicians need only a high school diploma. Mapping technicians require a little more education but can get by with an associate's degree in geomatics or a related field. The two jobs are a little different, though they are closely related.

Surveying technicians are usually out in the field, visiting sites and collecting data about the location. They then enter that data into a database. If you think that you would prefer to work mostly outdoors, then this is a good option for you. If you'd rather be indoors at a desk and behind a computer, then you may be better suited to being a mapping technician. Mapping technicians take the data the surveying team collected and use it to create maps of the terrain.

Most surveying and mapping technicians work for firms that provide engineering, surveying, and mapping services on a contract basis. Local governments also employ them in highway and planning departments. Both mapping technicians and surveying technicians need strong math skills, because they work with precise measurements. Usually, they'll also receive on-the-job training under a more experienced professional.

If you enjoy math but do not want to be tied to a desk, working as a surveying technician could suit you.

Going to math camp can be a great way to advance your math skills.

Career Insight:

Math Camp

If you're really interested in a career in math, you might want to try to participate in a math camp over the summer. There are many such camps offered around the United States and in Canada. Even local Parks and Recreation Departments often offer STEM and math day camps.

There are also particular organizations that run intensive, overnight math camps for talented math students. For example, Canada/USA Mathcamp welcomes high school students from the United States and Canada who are talented at math and interested in attending the five-week camp. Some universities, such as Stanford, offer programs too.

Some of the camps can be quite expensive but most offer financial aid options. There are also some camps designed for students who excel in math but whose families have low incomes. For example, Bridge to Enter Advanced Mathematics (BEAM) was established for this exact purpose and offers a free five-week day camp for sixth-graders in the New York City or Los Angeles areas who are exceptionally talented at math and come from low-income families. Camps such as BEAM aim to level the playing field for talented students. In years past, students from low-income families saw their own education suffer because they could not afford opportunities like these. BEAM and other organizations try to provide opportunities for these talented students.

A DAY IN THE LIFE OF An Architect

While every day is a little different in the life of an architect, these pages should give you an idea of what one day might look like.

8:00 a.m. You arrive at a job site for a morning meeting. Once a week you check in at each job site to make sure everything is on track. You trust the contractors on the job but that doesn't mean you shouldn't check in. In your years of experience, you've found that weekly check-ins can make all the difference to how a project works out. On this particular day, at this site, all is going well.

9:30 a.m. You arrive at the home of a potential client who wants to build an addition on his house. You go through the house with the client and listen to his ideas. You're not so sure they'll all work but there's time enough to discuss the particulars later. For now, you're just getting an idea of what the client wants and taking measurements so you can start to sketch out some visuals and plans for how you might be able to make his dream a reality.

11:00 a.m. You arrive at the office and answer emails and voicemail that have come in since you left yesterday. As usual, there are what seems like a million little questions to attend to but knowing all the details makes a big difference.

1:00 p.m. Having finally made it through all of the emails and voicemail, you grab a quick lunch to eat at your desk. You're anxious to get started sketching out some ideas for the client who wants the addition on his house.

Job site meetings are a good way to draw people back to your plans and make sure that they are still on track.

It feels good to present your plans to clients and get positive feedback from them.

2:00 p.m. You realize that you need the latest plans for the home, so you head down to the city offices to pull the original site plans. While you're there, you decide that you may as well get information about building codes for that city. You want to make sure your proposed design is workable before you put too much effort into it and take it much further.

3:30 p.m. You arrive back at the office in time for a meeting about a new renovation project for a historic building in the heart of downtown. It's an exciting project but one that will require a lot of careful work to preserve the building's historic status. You love this kind of project because you get to bring the old and new together.

4:45 p.m. Back at your desk, you answer the emails and voicemail that have piled up during the course of the day. You continue on the home renovation plans and sketches. As you've gone about things, you've been thinking of new ideas. You really hope the client likes your ideas as much as you do!

6:45 p.m. You've made a great start on the drawings but before you get too excited, you want to sleep on things and look at them tomorrow with fresh eyes.

Builder

Review and Check In

Now that you've learned about some of the builder roles in math, explore more about how you feel about each job. Use the questions below to help you assess how strongly you feel each role might suit you.

Architect

- You'll need to follow a client's brief to ensure the project is successful. What challenges would this bring and how would you deal with them?
- Following a brief also means keeping within a budget. Are you comfortable working to a budget? If not, how can you improve on this?

Urban Planner

- What do you think the challenges of the job might be?
- In this role, you would collaborate with a team of people. Would you enjoy that or would you find it difficult? Why do you say so?
- You would be responsible for managing budgets and schedules. Would you be happy with that responsibility?
- In this role, you would travel a lot. Would that suit you? How might that impact your lifestyle?

"When choosing a career, always consider the career outlook—does your chosen area have potential in the future?"

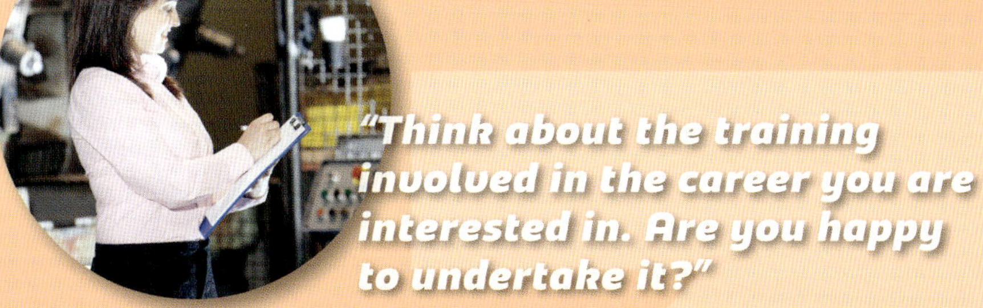

"Think about the training involved in the career you are interested in. Are you happy to undertake it?"

Mechanical Engineer

- What skills do you have that would make you well suited to this job?
- Often, mechanical engineers work long days and mostly in an office. Would this suit your personality? How might the long hours impact your lifestyle?

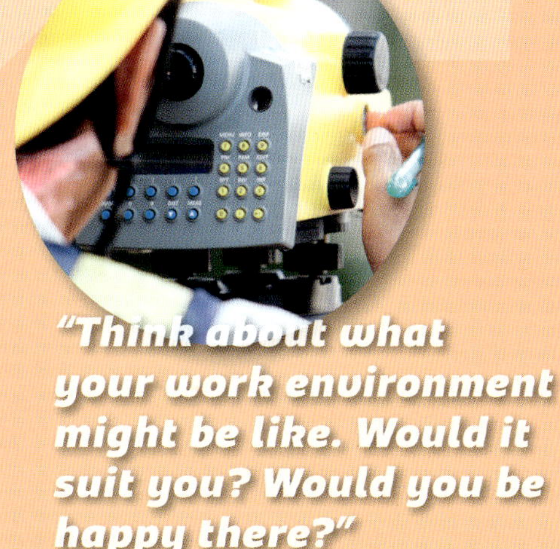

"Think about what your work environment might be like. Would it suit you? Would you be happy there?"

Surveyor or Mapping Technician

- As a surveyor you would spend a lot of time outdoors but a mapping technician may spend more time behind a computer screen. Which appeals more to you and why?
- What skills do you have to make you successful at these roles?

Research More

If you like the idea of working as a builder in math, but are not sure the jobs covered in this chapter suit you, here are some more builder roles you could explore.

Carpenter
Electrical Engineer
Formula 1 Engineer

CHAPTER FOUR

CREATOR ROLES IN MATH

Creative and math are two words that you don't often see together. But, actually, they go quite well together. Neuroscientists who study the brain have determined that an appreciation of beauty in math and beauty in art activate the same area of the brain. Brain scans have shown that this area of the brain becomes very active when the mind is working to integrate decision-making, emotion, and sensory experiences. In other words, the same part of the brain that appreciates a good math problem may also appreciate the beauty in art. For those creative people who are also skilled at math, there are a number of career opportunities in creative fields.

Animator

Animation used to be hand drawn, frame by frame, and it was tedious work that took forever. Today, most animation work is done on a computer using specific software. This speeds things up in a way—no more drawing the same characters over and over, with just a fraction of change in each frame to show movement. However, animation is still difficult work, and it involves a fair amount of math. This is because you have to tell the computer how to animate the character. The process involves understanding trigonometry to move and rotate the characters. It also involves algebra, creates the special effects that make images shine and sparkle, and calculus, which helps light the scenes.

Advanced animation software has completely revolutionized the way that animations are created today.

If you think you might enjoy a career in animation, why not try to improve your skills now by using some of the free graphics programs that help you develop your talent?

Animators can work at animation studios, such as Pixar or Disney, but they can also work for game development companies, such as Electronic Arts or Nintendo. Still others work freelance, using their talents for a variety of smaller clients. What software you use as an animator depends on where you're working. Some studios have their own proprietary software. For example, animators at Pixar use RenderMan, a 3-D rendering that was developed by the studio. People outside of Pixar can purchase a license to use the software too, but the software was actually developed by Pixar and for use by Pixar staff.

Most animators have a bachelor's degree in computer graphics, fine art, animation, or a similar field. They must also have a portfolio showing their strongest work. Given the technical nature of animation, this will be an online portfolio showing animation completed to date.

If you're creative and interested in animation but unsure where to start, there are some free software programs you can use to practice digital animation. For example, Blender is a free graphics program that you can use to practice your skills in 3-D animation, motion graphics, and visual effects.

Technical Writer

Another job in which creativity and math can collide is in technical writing. Technical writers write documents such as user manuals, guides, articles, and books on subjects such as technology. Their job is to make difficult material understandable for readers, which involves a certain amount of creativity.

Generally, technical writers are expected to have a college degree in English, journalism, communications, or the technical field for which they're writing. They need a degree that shows they have either a writing education or a technical education —but not necessarily both. In general, employers are willing to train on one or the other. For example, if you have a degree in advanced mathematics, you could be hired to write a math textbook, even if you're not the greatest writer. The editor can help polish up the rough spots in your writing. Alternatively, if you have a degree in English and are a talented writer, you may be hired to write a book on fun math tricks and games for kindergarten children.

No certification is required to be a technical writer, but if math is your field you might want to consider earning certification through the Society for Technical Communication (STC). It will likely open up some more job opportunities for you, as it proves you are competent to write about technical subjects such as math.

If you enjoy writing and find that your English skills are as strong as your math skills, you could combine both areas by working as a technical writer.

Working as a freelancer means that you can work from any location. Many people prefer the freedom this offers.

Career Insight:

Freelance vs. Employee

For some careers, such as technical writing, you have the option of working as an employee for a particular company, usually in an office, or working freelance, usually from home. Working for one particular company, as an employee, has some benefits. You earn a steady paycheck, usually get medical insurance paid for by your employer, you usually work in an office, so see other people each day, and you have predictability in your working life. Working freelance has other benefits: You often get the comfort of working from home and setting your own hours, to some extent. You have the freedom to work for multiple clients. You can set your rates, which are often higher than what you'd get paid from a single employer. You get to be your own boss. Working freelance from home can be lonely for those who like to be around other people, especially if you're the kind of person who likes to bounce ideas off other people. Providing your own medical insurance can be very expensive and you have to pay additional taxes if you're a freelancer. All of these factors are things to keep in mind when looking at a career that could be freelance or in-house.

Cartography is a skill that combines precise artistic talent with a skill for understanding and using numbers.

Cartographer

Another creative field where math comes into play is cartography—or the creation of maps. Maps used to be made and printed on paper but today, many maps are made on computers and used digitally on computers and smartphones. The job requirements are still the same for cartographers, though—they must use information, mathematical principles, knowledge of scaling, and models to create maps for navigation, geography, topography, weather forecasting, tourism, city planning, and much more.

Fashion design is not simply about choosing a pretty fabric! It involves finely tuned math skills.

Cartographers usually have a bachelor's degree in cartography, geography, surveying, geomatics, or a related field. In some states, cartographers must also be licensed as surveyors but that's not the case in every state. There are also a few certifications available for cartographers who are interested in earning a certification in the field. Having a certification will open up more doors when it comes to finding a job in this field.

Career Insight:
High Fashion

It might surprise you to learn that there's a need for strong math skills in the unlikely field of fashion design. The human body is a study in geometry—it's full of different shapes and angles. When you put fabric over those shapes and angles, unless you know how shapes lie on one another the fabric will not drape and fall in the way you intended. Try it: Cut a hole in an old bedsheet and put your head through it. You'll discover that when you put it on, it pretty much looks like an old bedsheet is hanging on you! There's no contour and no pleasing design.

Fashion designers take two-dimensional (2-D) fabric and lay it on a 3-D model to determine how best to work with the fabric and make an attractive piece of clothing. This may be done on a mannequin or a live model, or it may be done on computer or even hand drawn. Either way, it involves understanding enough math to know how fabric will look on a body.

Designers also need to know math to scale the size of the garment. The sample garment will be in a particular size but it will need to be made bigger and smaller to fit people of other sizes. That requires the use of math and knowledge of scaling measurements.

And then of course there's the math needed to determine pricing. If materials to make a shirt cost $15 and it took three hours of time for you to make the shirt, you obviously don't want to price it at $20. You'd make no profit! After subtracting the cost of the material, you'd be left with $5 to cover the three hours of time you spent making it. That's not even a legal wage. So you'll certainly use your math skills when determining pricing. (And if you don't, you might find yourself broke!)

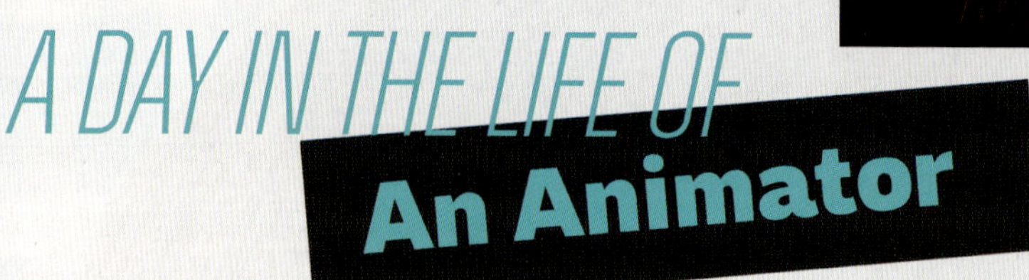

A DAY IN THE LIFE OF An Animator

Many animators work for studios. Although every day at work in a studio looks a little different, this "day in the life" describes a typical day as an animator for a major animation studio.

8:30 a.m. You arrive at the office and answer any emails before you fully start your day.

9:00 a.m. It's time for dailies! You have to show the series of shots you created yesterday to the entire animation team for critique. When they like the shots, it's great–you feel like you're on top of the world but sometimes, there's a lot of constructive criticism. It's great feedback but it means a lot of rework for you.

10:30 a.m. Dailies are over and done, and today there were only a few minor suggestions to your shots from yesterday. You set to work cleaning those up.

12:00 p.m. You grab a quick lunch at the cafeteria with some of your fellow animators. It's good to get a short break from the screen and it's always good to chat to your colleagues, especially when you're stuck on something. One of your colleagues has a great idea for the next lot of shots so you're grateful.

The feedback from dailies can mean just a few tweaks to projects or a complete overhaul of your work.

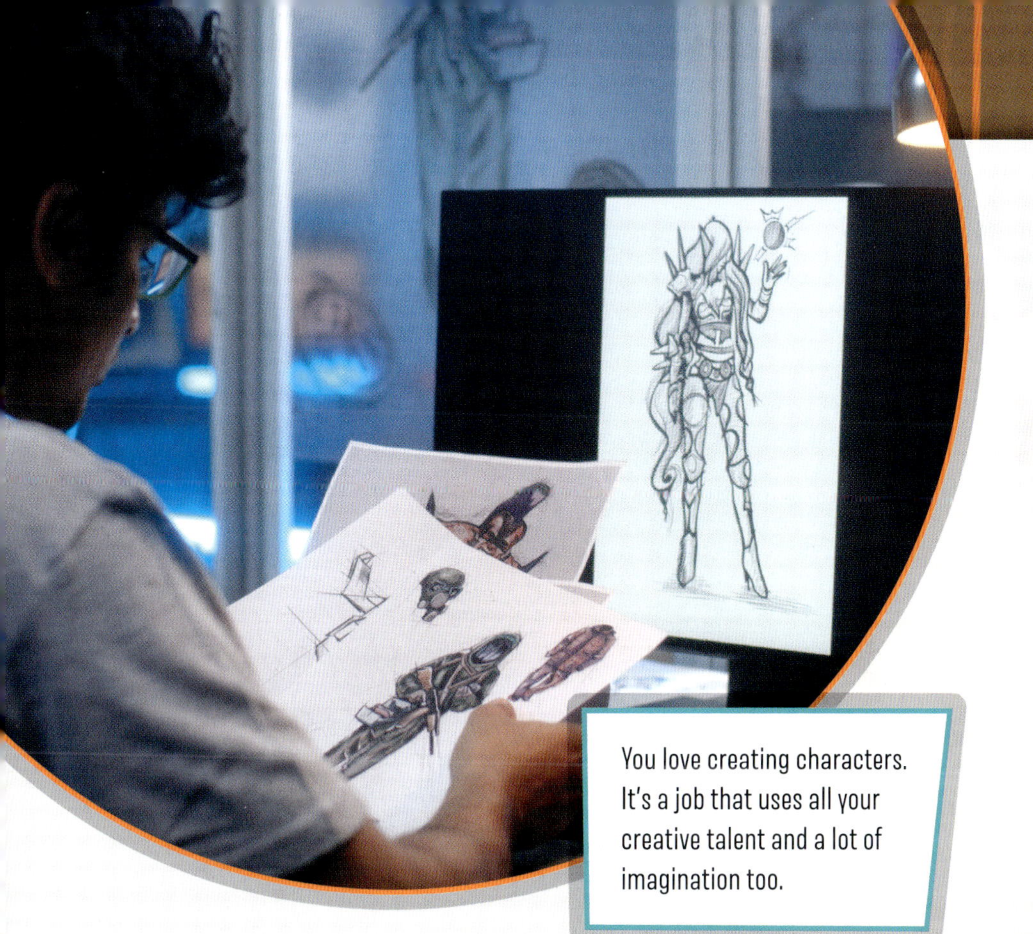

You love creating characters. It's a job that uses all your creative talent and a lot of imagination too.

1:00 p.m. You get to work on the shots for today. You're working with the animation studio's proprietary 3-D animation software, which you're very comfortable with. It's easy to use, but the detail in the work still requires a lot of focus.

3:00 p.m. You tear yourself away from your animation work to lead a segment of a tour. School groups often tour the studio and you're in charge of leading a 30-minute part of the tour focusing on the studio's history. It's a fun way to break up your day, though it is hard to tear yourself away from work when you're making so much progress.

3:30 p.m. The tour went well but it's back to work on the day's shots. They're coming along well, though the character you're animating has long, thick hair that is difficult to animate in motion scenes!

5:30 p.m. You've wrapped up your studio work for the day, so technically you can go home. But the studio encourages employees to work on their own independent projects, too, so you decide to stay at work for a couple more hours to work on an animated short film you're hoping to finish by the end of the year so you can enter it into next year's short-film festivals!

Creator

Review and Check In

Now that you've learned about some of the creator roles in math, explore more about how you feel about each job. Use the questions below to help you assess how strongly you feel each role might suit you.

Animator

- Why would you enjoy combining your artistic talent with computer skills to create an animation?
- Which field of animation would you like to work in? How are your skills suited to this field?
- Some animators work as freelancers. Would you like to be your own boss or work for someone else? Why?
- If you work for a large animation studio, you will need to be a team player. What makes you the kind of person who likes to collaborate with others?

Technical Writer

- Being a writer can be a lonely job. Would this be challenging? If so, how would you deal with these challenges?
- Most writers are their own bosses. What skills do you have to enable you to be successful at this?
- Part of the writing process means that your work will be edited. Would you be happy to have your work edited? If not, how can you adapt to this?
- Are you good at keeping to deadlines? If this is not your strength, how can you be better at it?

"Ask yourself, 'Will I enjoy doing the job every day?'"

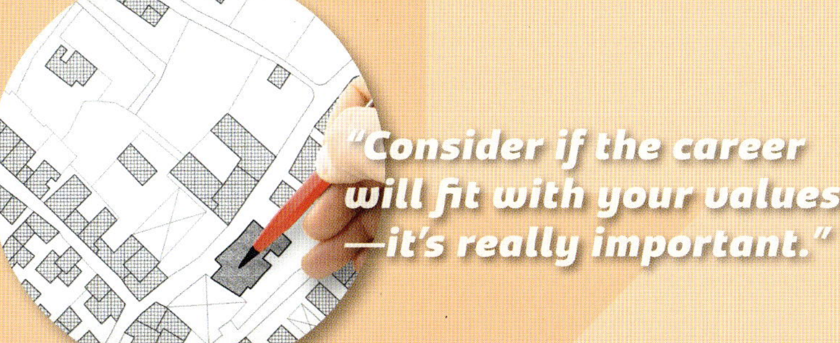

"Consider if the career will fit with your values—it's really important."

Cartographer

- What would you find rewarding about the job? What might the challenges be? How would you deal with those?
- Are there certain parts of the job that you like more than others?
- On larger projects you would need to work as part of a team. What skills could you bring to the team to ensure the success of the project?
- Your job would be a mixture of field work and office work. Why do you think you would like this mixture and flexibility in your job?
- Would you be prepared to study further to get your certification?

"Think about the financial side of your chosen career—will it suit your lifestyle plans for the future?"

Research More

If you like the idea of working as a creator in math, but are not sure the jobs covered in this chapter suit you, here are some more creator roles you could explore.

Architectural Technologist
Marine Architect
Music Tour Manager

CHAPTER FIVE

ORGANIZER ROLES IN MATH

If you love having things in order and chaos is your enemy, you're an organizer! Maybe that's partly why math appeals to you. There are rules and it's predictable, with very few surprises. Let's discover some organizer roles you could do.

Accountant

If you're an organized person and you love numbers, accounting might just be your dream career. It's all about balancing budgets and keeping profits and expenses organized. Filing income taxes truly is the ultimate exercise in number organization!

There are a number of different subfields in accounting. Public accountants perform accounting and tax-preparation duties, among other tasks. They either work for a public accounting firm or they have their own accounting business. Either way, most choose to take the certification exam necessary to become a Certified Public Accountant (CPA).

Management accountants are also called cost, corporate, industrial, or private accountants. They analyze the financial data for organizations, not for individual people. Their data work is specifically for use by the business's managers; it's not part of the public accounting for the business. They may manage the business's investments, work on budgets, and evaluate the company's financial performance.

Accountants are usually super-organized with an eye for meticulous detail. They enjoy keeping accounts in order and tying up any annoying loose ends.

Bookkeepers use their math skills to keep company finances organized. They are skilled at using spreadsheets and computing data.

Government accountants work with records from government agencies. They also audit small businesses and individuals. During an audit, a government accountant will pay careful attention to the financial information and tax records of the individual or business and make sure taxes have been paid appropriately.

Accountants must have at least a bachelor's degree in accounting or a similar field, such as business. Further education may be required for higher-level management-type positions. However, depending on the accounting firm, there may be room for an individual with a bachelor's degree to move up, especially if that person is also a CPA. To become a CPA, you must pass a national exam and meet the requirements of the state in which you're working. If your job will require you to file any reports with the government's Securities and Exchange Commission (SEC), you must be a CPA. However, some accountants choose to become CPAs even if they aren't working with the SEC, to improve job opportunities.

Bookkeeper or Accounting Clerk

If you're not ready to get a four-year degree in accounting but you're interested in the field, you can try being a bookkeeper or accounting clerk. Bookkeepers and accounting clerks produce financial records. Most of their training is typically on the job, and they need only a high school diploma to be hired. However, some businesses would prefer to hire a bookkeeper or accounting clerk who at least has some education beyond high school, such as accounting, math, or computer courses taken at a community college.

Checking company reports is an important job—the report shows how a company is performing financially.

Becoming a Bookkeeper

Bookkeeper and accounting clerks generally work under accountants, although some smaller businesses do not have accountants on staff, so the bookkeeper may not always report to an accountant.

Both bookkeepers and accounting clerks enter financial transaction data into the appropriate software. This includes costs and income in all forms—by cash, check, credit card, voucher, and so on. They produce balance sheets (which compare cost and income) and income statements. They also double-check financial accuracy in all company reports.

Much like there's a CPA certification for accountants, there's also a Certified Bookkeeper (CB) certification for bookkeepers who have at least two years of bookkeeping experience, have passed a four-part certification exam, and have demonstrated that they can comply with a code of ethics.

So, if you think you want to try being a bookkeeper or accounting clerk, you might want to set a long-term goal of someday earning your accounting degree. That way, you'll have more job opportunities in the future.

If you enjoy analyzing mystery in numbers, forensic accountancy could be your perfect job.

Career Insight:
Accounting and the Law

Some people assume accounting is not a very exciting career. There's a stereotype that accountants are quiet, serious, nerdy sorts who are happy to spend their days buried in lists of numbers! Part of that may be true: Accountants usually do enjoy spending their days thinking about numbers but the rest of the stereotype is not quite true. Accounting appeals to all types of people who enjoy working with numbers and if you're looking for a little excitement in an accounting career, you might think about becoming a forensic accountant.

The word forensic refers to using scientific methods to investigate crime. So forensic accounting is using the methods and practices of accounting to investigate crime. Usually, this involves property crimes, such as fraud and embezzlement. People who have committed crimes like this typically go to great lengths to hide the crime. After all, no one wants to go to prison! So forensic accountants have to be careful detectives and try to find hidden evidence of a crime. They must go through a person's or company's financial records with a fine-toothed comb to see whether anything indicates evidence of a crime.

Sometimes forensic accountants have to go beyond the books and actually interview people involved in the suspected crime. In such cases, they're almost more detective than accountant—though of course the crime revolves around the numbers. And at times, forensic accountants have to testify in court about what they've found in their investigation.

If you're interested in a career in forensic accounting, you'll need a bachelor's degree in accounting, and you'll need to pass the examination to become a CPA. There are also several certifications available for forensic accountants.

Stockbroker

Stockbrokers invest in the stock market for their clients. Their clients are sometimes individuals but other times they are businesses.

The stock market is where shares of stock are sold. Shares of stock are like pieces of businesses. When a business is privately owned, an individual or a small group of individuals owns the company. For example, Publix Supermarkets is a privately owned company. The Jenkins family (who founded the company) and Publix employees own Publix. The owners all own small pieces of the company called shares. No one from the general public can own Publix shares—only the Jenkins family and past and present employees of Publix can.

For a publicly owned business, however, shares can be bought and sold on the stock market. For example, Apple and Microsoft are publicly owned. Anyone can buy shares in them on the stock market. And if they wanted to do so, they might have a stockbroker help them. People can also buy stocks on their own but stockbrokers know the stock market inside and out, and can often give good advice about what's smart to buy and what isn't.

Stockbrokers must keep a watchful eye on how stocks and shares are performing and react quickly as the market changes.

Career Insight:
Turning Down the Heat

The world of stocks and stockbrokers can be pretty cutthroat. If working with stocks and shares interests you but you aren't sure you want to work at a major stock brokerage, you might want to become a personal financial advisor. This role utilizes your math skills, but is less stressful.

Financial advisors work with individuals on how to invest their money, how to save for their children's college educations, how to save for their retirement, and more. They keep up with the stock market to advise their clients on investments, but they aren't working on the fast-paced floor of the stock exchange, where decisions are made minute to minute.

Similarly, to do this job, you'll need a bachelor's degree in economics, accounting, business, finance, or a similar field. As you work for more people, you'll need to develop your portfolio of clients—people like to know what you've done for others before you touch their money.

To become a stockbroker, you need a bachelor's degree in economics, accounting, business, finance, or a similar field. Technically, you don't have to have the bachelor's degree, since anyone can buy and sell stocks but chances are, no one will hire you as a stockbroker unless you have that degree. They want to know you have the education behind you to make you a good stockbroker.

You can also earn a master's degree in a related field if you want to move up to higher or more managerial positions. But an entry-level position as a stockbroker really only requires a bachelor's degree. To be a working stockbroker, you also need to pass a licensing exam.

Financial advisors have the responsibility of helping people figure out how to best invest their money so that it helps them realize future dreams and plans.

A DAY IN THE LIFE OF An Accountant

As in most careers, no two days ever look the same. For a CPA, there are certain tasks that are done routinely at specific times of the month. Here's a glimpse into what a typical day might look like.

9:00 a.m. You arrive at your office and start going through emails that have come in since you left yesterday. The great thing about having your own accounting firm is that you can set your working hours. You like jogging before work so your firm opens at 9:00 a.m.

9:30 a.m. The first of many phone calls interrupts your email answering. It's a welcome interruption, though, because phone calls usually mean paying work. This case is no exception. It's an individual who wants to talk to you about creating a business plan for a small business he plans to start. You set a meeting for the afternoon—you happen to have an opening in your schedule today.

9:45 a.m. You get back to work answering email. One of your clients is struggling with a loan application, so you decide to call and talk through it on the phone with her. It's much

You must keep ahead of all the latest developments in tax law so you can keep your clients informed.

Analyzing proposed business plans requires that you pay attention to the detail so that you can flag up any issues to your clients.

easier to spend 5 minutes on the phone. You have a monthly meeting at the Chamber of Commerce. It's nearby so you can just walk over to the meeting.

11:00 a.m. Your meeting over, you head back to the office just in time for an 11:00 a.m. meeting with another client who is considering expanding her business. Her plan is a good one but the funding is tricky. You spend longer than you intend talking through the options with her but she leaves with a lot of good information.

12:30 p.m. You grab a muffin from the coffee cart downstairs. No time for a full lunch today, since your last meeting ran over.

1:00 p.m. A client arrives for his appointment. He wants to talk to you about his business plan. He has a solid idea but no idea of the tax implications of what he intends to create. You spend a couple of hours talking about funding and tax structures with him. Like your earlier client, he leaves armed with good information and a way forward.

3:00 p.m. It's payroll day, so you begin processing payroll for the three other employees in your office. Since your firm is quite small, it doesn't take long.

3:30 p.m. As you're on a roll with bookkeeping, you decide to complete bookkeeping tasks that have cropped up for a few clients. You're committed to doing them within two days of when the clients send in the receipts, which means they're due tomorrow.

4:30 p.m. You decide to close up the office early, once you finish the bookkeeping tasks. Tax season is coming up, which will mean longer than usual hours for everyone in the office. You might as well let everyone go home early while you have the chance.

Organizer

Review and Check In

Now that you've learned about some of the organizer roles in math, explore more about how you feel about each job. Use the questions below to help you assess how strongly you feel each role might suit you.

Accountant

- You would need to understand tax laws and regulations to do this job well. Why would you find that interesting?
- You will need to work to the same deadlines each month. Would you enjoy being methodical like this? How would you make sure these deadlines are met?
- Are you prepared to study further to get your CPA qualification?

Bookkeeper or Accounting Clerk

- Being organized and meticulous are key to success in this role. What else would make you good at it?
- You may need to wait on others to complete their work before you can do yours. Would this be a problem for you? Why or why not?
- What would you find challenging about this role?

"Always keep an eye on the future and how your career might change. For example, will technological advancements change some aspects of your job?"

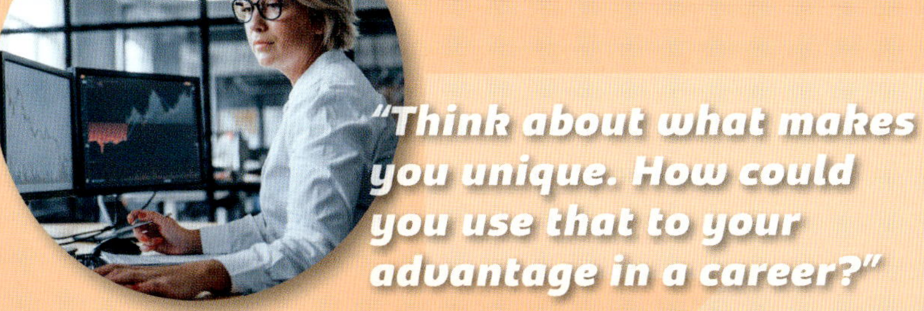

"Think about what makes you unique. How could you use that to your advantage in a career?"

Stockbroker

- How would you deal with the pressure and stress of being a stockbroker?
- You could be your own boss or work in a large firm. What would you prefer and why would you prefer that?
- You would need to think "outside the box" to be a broker. Is this something that you would enjoy doing?

"Never rush into career decisions. Always take your time, and work through all your options."

A DAY IN THE LIFE OF An Accountant

- Review the Day in the Life feature. Did anything about the day surprise you? What was that?
- What do you think the challenges were, and how might you deal with them?
- What aspects of the role do you think you would enjoy?
- Look at the structure of the day and the working hours. Would you be happy with that structure? How might it affect your lifestyle?

Research More

If you like the idea of working as an organizer in math, but are not sure the jobs covered in this chapter suit you, here are some more organizer roles you could explore.

Accounts Assistant
Pharmacy Technician
Supply Chain Manager

CHAPTER SIX

THINKER ROLES IN MATH

Thinkers like to consider facts carefully and analyze them before making a decision. Math is all about thought and analysis, so you're in luck—there are many math careers well suited to your career personality type.

Actuary

Actuaries analyze the financial costs of risks and uncertainty. They often work in the insurance industry. People buy insurance to pay out if certain events happen. For example, they buy car insurance that will pay for car repairs and medical bills if they are in a car accident. People buy health insurance to help pay their bills when they are sick or injured.

Insurance companies don't just hand out insurance to everyone who asks, though. If they did, they'd go broke! You might pay $100 per month for car insurance, for example, but if you get in a car accident and the damages are $100,000, you can see that the insurance company is going to pay out a lot more money than they ever took in from you! Unless, of course,

If you become an actuary, your role will be to figure out how much risk is associated with an individual, for example, how likely they are to have a motor accident.

Insurance is big business involving many millions of dollars. Managing insurance funds requires logical thinking and an analytical mind.

you paid your $100 per month for 1,000 months, at which time the insurance company would have collected $100,000 from you. But that's more than 83 years of paying insurance, and since most people don't start paying car insurance until they're at least 16 years old, this situation is quite unlikely.

The above scenario is exactly the kind of thing actuaries think about. They use math, probability, and data analysis every day to determine how likely it is that an event will occur that will result in the insurance company paying out a large sum of money. If an applicant for car insurance has had three accidents in their first two years of driving, the insurance company actuaries will recognize that the person is quite likely to have another accident, which means the insurance company will likely have to pay out a claim. In that situation, they will either deny the person insurance coverage or will offer coverage but at a very high price. Like any other business, insurance companies are in business for one reason—to make money and not to pay out every possible claim.

Actuaries must have a bachelor's degree, usually in mathematics, actuarial science, statistics, or a similar field. There are several different fields for actuaries, including health, life, pension, and casualty. These different types of actuaries work in insurance fields covering auto, homeowners, medical malpractice, workers' compensation, life, health, retirement, investments, and finance. There are two different certifications actuaries can earn, depending on which of these fields they want to specialize in.

Cryptanalyst and Cryptographer

Cryptanalysts are also known as codebreakers. They analyze hidden or encrypted information and translate it into understandable form. Cryptographers do the exact opposite: They encrypt sensitive information that needs to be transmitted without everyone being able to read it.

Cryptanalysts and cryptographers work with sensitive financial data or matters of national security. For example, terrorist groups may send encrypted information about planned attacks, and cryptanalysts can help decrypt that information and alert the appropriate authorities so they can take action and prevent the attack.

Cryptographers may help companies encrypt sensitive personal or financial information so that customers' personal information is not at risk. In recent years, major companies such as Target have experienced data breaches, in which customer information was stolen. To prevent these breaches, companies may use cryptographers to help protect sensitive information.

In the online age, cryptographers and cryptanalysts are in greater demand because so much sensitive information is stored and transmitted over the Internet. There are specific bachelor's and master's degrees available in cybersecurity, for those interested in working specifically in online cryptanalysis or cryptography. There are also degrees in computer forensics and digital investigation, as well as degrees in information technology (IT) with a specialization in networking and security. All of these are reasonable paths to follow if you're interested in a career in cryptanalysis or cryptography. You can also obtain a bachelor's degree in a more general field, such as mathematics or computer science but you might want to double major in both math and computer science to make yourself a competitive applicant for jobs in the field.

Cryptanalysis and cryptography are growing areas of work as cyberattacks and cybercrime become ever-increasing problems.

This is the infamous Enigma machine that was used to crack codes sent by the Germans during World War II.

Career Insight:
Women in Codebreaking

Cryptanalysis was a big field during World War II. Nazi forces and the Japanese sent encoded messages about attacks against the Allies. Women played a big role in codebreaking during World War II. For example, Elizebeth Smith Friedman was married to William Friedman, a cryptographer for the Army. At that time, women mostly worked in factories, taking over when men went to war, or as secretaries. Friedman was different, though. She was a talented codebreaker and her work was often recognized as a joint effort with her husband. In reality much of her work was her own. She was instrumental in breaking code about the Pearl Harbor attack that was being transmitted by Velvalee Dickinson, who was known as the Doll Woman, and she also helped break the exceptionally difficult enigma code transmitted by the Germans in World War II.

Friedman's background was not in mathematics. Interestingly, she majored in English literature in college and also studied Latin, Greek, and German. She just happened to have a real knack for breaking code. Today, she is recognized as the United States' first female cryptanalyst.

If you love staring at mathematical problems and working them through, it could be that you'd find your dream job as a mathematician or statistician.

Mathematician and Statistician

If you really want to eat, sleep, and breathe math, you might want to consider becoming a mathematician or statistician. Mathematicians and statisticians take data and apply mathematical theories and statistical techniques to come up with solutions to problems in many different fields.

Sometimes mathematicians and statisticians develop new mathematical rules and theories, and sometimes they use long-existing ones to prove new work. Either way, they apply techniques to help solve problems in business, science, engineering, or other fields. They sometimes design surveys or experiments to help them gather the data they need to work on the problem. And when they have gathered the data they need, they interpret it and draw conclusions about the best solution. They then prepare reports that communicate their analyses and results. Some Mathematicians and statisticians also manage databases, and once the data they have gathered is analyzed, they interpret the results, draw conclusions, and make reports for clients and upper management.

Mathematicians and statisticians work in many fields. Some are obvious fields related to numbers, such as finance or business, but they work in areas you might not expect, too. For example, when pharmaceutical companies develop new drugs, they have to run countless tests and studies to determine the drug's effectiveness and any possible negative effects. Mathematicians and statisticians look at the data collected from those tests and studies and draw conclusions about the drug, its effectiveness, and its potential danger.

To begin a career as a mathematician or statistician, as well as having strong analytical, organizational, and problem-solving skills, you'll need at least a bachelor's degree in math, statistics or a related field. Many companies and organizations will expect you to have a master's degree.

The movie *Hidden Figures* highlighted the until then little-known story of the three women whose mathematical brilliance helped make space travel possible.

Career Insight:
Hidden Math Talents

The movie *Hidden Figures* introduced the world to a group of women with incredible talent in mathematics who made space travel possible but who for years, received hardly any recognition for their efforts and talent. Katherine Johnson, Mary Jackson, and Dorothy Vaughan were three African American women who worked for National Aeronautics and Space Administration (NASA) as "human computers," performing difficult mathematical computations that would ensure astronaut John Glenn's 1962 space mission was a success. The three women were just three of many women who worked in mathematics and engineering positions at NASA at that time but none of them got much attention. In that era, women who did work outside the home were usually teachers, secretaries or nurses, not mathematicians and engineers. The women of NASA were brilliant minds who have in recent years been recognized for their incredible contribution to the space program.

A DAY IN THE LIFE OF A Cryptographer

Being a cryptographer is really interesting and like many careers, each day is different. Here's an idea of what a day might look like.

8:00 a.m. You arrive at the office, ready for the day. Your work can be tiring so you like the early start while your brain is fresh. You address any emails and messages that have come in since you left yesterday.

8:30 a.m. It's time for your morning meeting with the team. As a team, you've been tasked with designing a secure operating system. It's a huge project that will take a long time and the effort of many people, so this meeting is the first of many you'll have as the project starts to take form.

10:00 a.m. You head back to your desk and start jotting down some notes from the meeting. Because you have some past experience with white-box cryptography (WBC), the project lead wants you to explore the ways you can use WBC for the project to minimize security risks. WBC uses math to blend together app code and keys to secure cryptographic operations. This prevents those keys from being found or extracted from the app.

11:00 a.m. You set aside your white-box research for a while and turn your focus to another project you're working on: Writing a compiler for a minimally secure central processing unit (CPU) as well as a medium-security CPU.

If you are working on a big project, putting in extra hours is often a requirement, especially if the project is time-sensitive and needs urgent action.

You never tire of staring at data–finding anomalies is satisfying, and that's what you are paid to do after all!

12:00 p.m. Your brain is spinning so you take a break from your desk and get some lunch. It's been a long morning of meetings, research, and development. It's fun work, but definitely requires brain power. And a lot of it!

1:00 p.m. Back at your desk, you start working on an attack of an existing system for one of the company's clients. One of your fun tasks as a cryptographer is trying to break into existing systems, so that you can address security flaws and make the system more secure.

3:00 p.m. You switch gears once again to another task you're working on: designing a secure communication protocol for a specific chip. You focus on the task.

4:30 p.m. You have a bit of time before you head home so you go back to researching WBC. It's a big project and you want to stay ahead of the game. Plus, you enjoy it, too!

Thinker

Review and Check In

Now that you've learned about some of the thinker roles in math, explore more about how you feel about each job. Use the questions below to help you assess how strongly you feel each role might suit you.

Actuary

- Actuaries typically work mainly in the insurance industry or in banking. Which do you have a preference for, and why?
- What skills would you bring to this job and what would you find challenging about it? How would you deal with the challenges?
- Do you have the communication skills to explain the risks to clients?

Cryptanalyst and Cryptographer

- Why does this job appeal to you? Are there things you may not find as interesting?
- As part of your job, you'd be working with sensitive financial data. Would that be a challenge for you?

"Close your eyes and picture yourself doing your ideal job in the future. What type of workplace setting are you in and what are you doing? Keep that vision in mind as you research careers."

"Write down your top five strengths and weaknesses on a piece of paper. Keep it to hand and use it to help you work through career options."

Mathematician and Statistician

- What aspects of the job do you think you would enjoy? Why do you think so?
- Are there parts of this role that you would find challenging? Would you enjoy those challenges?
- If you were part of a team, would you prefer this to working alone? Why?
- There are many different fields where you would need to use your skills. Which field is your preference and what makes that field appealing?

"Once you know what career you'd like to follow, make a flowchart that shows all the steps you need to take to get to that career —then follow it!"

Research More

If you like the idea of working as a thinker in math, but are not sure the jobs covered in this chapter suit you, here are some more thinker roles you could explore.

E-Commerce Business Manager
Economist
Cosmologist

WHAT NEXT?–YOUR CAREER CHECKLIST

If you have come to the end of this book and are ready to start making a career in math a reality, follow the checklist on these pages to kick-start your future.

Start at School

Naturally, all roles in math will require strong skills with figures, so focus on your math class at school. Some of the math-based roles explored in this book require a good grasp of science, so make sure you pay attention in those classes as well. If your school offers computer science and IT courses as options, make sure you sign up for them. In those courses, you'll become familiar with many different aspects of the type of skills you'll need in builder, helper, or organizer roles.

Creator roles in math will allow you to mix your creative talents with your math skills. Studying visual arts and computer science at school will help you focus on these two areas. Also focus on your science and English classes if you wish to get into technical writing. And get involved in any appropriate clubs at your school or outside of it–gaining experience now will help set you up for a future career.

Talk to a Guidance Counselor

If your guidance counselor offers career advice, take it! Guidance counselors will be able to help you explore a lot of different options and talk more about whether a role in math could suit you. They will also be able to advise you on further education courses you could pursue after school or in-job training that could suit you.

Get Connected

Your parents and parents' friends may have great contacts in math-related fields. They may be able to put you in touch with an employer, so that you can talk to them directly about roles in math that interest you. For example, a parent or friend could connect you with an architecture company, so that you can talk to the people who work there about the roles they carry out.

Some employers are happy to have students meet with them in person for informational interviews, through which you can find out what people in math roles do. You may even be able to shadow someone who works in math. For example, you could spend a day or two at work with an architect or a surveyor, finding out more about what they do and how they do it. That's a really great way to discover if you would like to do their jobs in the future.

The best way to find out if a career is really going to suit you is to try it out! Many businesses offer internships, which are unpaid temporary jobs in which people can get work experience. Internships usually run over the school summer break, to give young people the best chance of trying out a role over a longer period of time. Most employers only take on interns who are 16 years or older, but sometimes internships have been offered to students as young as 14 years old. An internship will look great on your resume when you come to apply for jobs in the future, and some internships even come with an academic credit if they are built into an educational program. Ask your guidance counselor if such an educational program is available at your school.

Make Your Summer Count

Getting a summer job is a great way to get experience, even if it is at the very bottom of the ladder. Working in any math-related company, no matter how basic your role is to start with, will give you valuable experience that will help you in the future. For example, just through helping out in an architecture company with their office admin, you will discover if a career in this area might suit you.

You can never do enough research when it comes to planning a future career, so explore as many resources as you can to find out more. Start by taking a look at the resources on page 63 of this book to learn more. For example, the U.S. BLS offers some great resources that can help you learn more about possible roles in math. Check out in particular their Occupational Outlook Handbook section.

As you research, remember there is never a right or wrong in making choices, and you can always change your mind. Keep flexible, be positive about your future, and have fun choosing your perfect STEM career.

GLOSSARY

algebra a branch of mathematics dealing with equations and functions

Allies a group of nations in World War II that fought together to defeat the Nazis

analysis detailed examination of something, usually for the purpose of drawing conclusions

associate's degree a two-year degree awarded by a community college

blueprints design plans for buildings or other structures

brokerage a company that buys or sells assets or goods for clients

casualty a type of insurance that deals in accidents or disasters

cholera an infectious disease of the small intestine, generally contracted from unclean water sources

Crimean War a war between 1853 and 1856 that took place between Russia and an alliance of Great Britain, France, Turkey, and Sardinia

embezzlement theft or misuse of money belonging to one's clients or employer

encrypted encoded

fraud deception that is intended to result in financial or personal gain

freelance working for multiple companies or people as a self-employed individual, rather than being employed by just one organization

genetics the study of inherited characteristics

malpractice improper, negligent, or illegal professional activities, usually by lawyers, medical professionals, or public officials

meteorology the branch of science dealing with the atmosphere, including weather forecasting

pension regular payments made during a person's retirement from an account that the person has acquired over the course of their working life

PhD doctor of philosophy degree. PhD degrees are available in many subjects

pioneer the first person to do something

probability the likelihood of something occurring

scaling representing an object in greater or smaller proportional dimensions, according to a common scale

software the programs and operating system used by a computer

topography a detailed map representation of the natural and artificial features of an area

typhoid an infectious disease that causes severe intestinal problems and a rash

workers' compensation medical benefits and income paid to a person who is injured at work

FIND OUT MORE

Books

Darling, David, and Agnijo Banerjee. *Weird Math: A Teenage Genius and His Teacher Reveal the Strange Connections Between Math and Everyday Life*. Basic Books, 2018.

Hynson, Colin. *Dream Jobs in Engineering* (Cutting-Edge Careers in STEM). Crabtree Publishing Company, 2017.

Resler, T.J. *How Things Work: Then and Now*. National Geographic Kids, 2018.

Websites

Take a look at the BLS site for more careers guidance:
www.bls.gov/k12/students/careers/how-can-bls-help-me-explore-careers.htm

Check out the BLS Occupational Outlook Handbook to find out more about different jobs and the qualifications you need for them:
www.bls.gov/ooh

If your love of math has led you to an interest in coding, this is the site for you. Discover tons of free lessons on how to code in common and lesser-known programming languages:
https://code.org

If you love math but are a more hands-on learner, this is the site for you. It specializes in hands-on, kinesthetic learning of math concepts:
http://nlvm.usu.edu/en/nav/grade_g_4.html

This website connects teens with experiential learning opportunities, including summer programs, community service opportunities, and other programs:
www.teenlife.com

Publisher's note to educators and parents:
All the websites featured above have been carefully reviewed to ensure that they are suitable for students. However, many websites change often, and we cannot guarantee that a site's future contents will continue to meet our high standards of educational value. Please be advised that students should be closely monitored whenever they access the Internet.

INDEX

accountants 40, 41, 46–47, 48, 49
actuaries 50, 51, 58
air traffic controllers 12, 13, 19
animators 30, 36–37, 38
architects and architecture 4, 9, 20, 21, 26–27, 28, 60, 61
biostatisticians 4, 10, 18
bookkeepers/accounting clerks 41, 42, 48
Bridge to Enter Advanced Mathematics (BEAM) 25
budgets 28, 40
builder personality and roles 5, 6, 8, 20–21, 22, 23, 24, 25, 26, 27
cartographers 34, 39
certification 13, 24, 33, 54
climatologists 14, 19
clubs at school 60
college 4, 22, 24
community college 41
computer science courses at school 60
considering challenges of the job 18, 19, 28, 29, 38, 48, 49, 58
considering rewards of the job 18, 19, 39, 49, 58, 59
considering working as part of a team 18, 28, 38, 39, 59
considering your finances 33
considering your lifestyle 19, 28, 29, 33, 38, 39, 49
considering your skills 18, 19, 29, 38, 39, 48, 58, 59
considering your work environment 5, 29, 33, 39
creator personality and roles 5, 30–31, 32, 33, 34, 35, 36, 37, 38, 39
cryptanalysts and cryptographers 52, 56–57, 58
degrees 4, 11, 12, 13, 21, 22, 23, 31, 32, 34, 41, 42, 43, 45, 52, 54
diplomas 24, 41
Disney 31
Electronic Arts 31
English classes at school 60
epidemiologists 11, 18
exploring options 4, 9, 55
fashion designers 34, 35
flowcharts 5, 8, 6–7, 59
forensic accountants 43
forensic science technicians 13
guidance counselors 60, 61
helper personality and roles 5, 6, 10–11, 12, 13, 14, 15, 16, 17, 18, 19
high school 12, 16, 17, 24, 25, 41, 60
informational interviews 61
internships 21, 61
IT courses at school 60
licenses 21
making connections 60
math classes at school 60
math teachers 12, 16–17, 19
mathematicians and statisticians 54, 59
mechanical engineers 22, 29
middle school 12
Nintendo 31
nurses 4, 15
on-the-job training 21, 24, 41, 60
online tuition 31
organizer personality and roles 5, 7, 40–41, 42, 43, 44, 45, 46, 47, 48, 49, 50, 51, 52, 53
personal financial advisors 45
Pixar 31
portfolios 31
research 9, 19, 29, 39, 49, 61
resumes 61
salaries 33
schedules 28, 38
science classes at school 60
self-employment 31, 33, 38
Society for Technical Communication (STC) 32
stockbrokers 44, 49
summer jobs 61
summer math camps 25
surveyors or mapping technicians 24, 29
switching careers 9
taxes 33, 40, 41, 46, 47, 48
technical writers 32
Tesla 23
urban planners 4, 21, 22, 28
visual arts at school 60
women in codebreaking 53
women in engineering 22
women in math 15, 23, 53, 55
working from home 33
working hours 16–17, 26–27, 29, 36–37, 39, 46–47, 49, 56–57

About the Author

Cathleen Small has written many books for young people on a wide variety of topics. In writing this book she has learned the value of personality testing, research, and considering many options when exploring a future career.